Josineto de Souza Alves
Simone Alves Silva
Francisco Souza Fadigas

Phytoremediation of soils contaminated by lead and cadmium

Josineto de Souza Alves
Simone Alves Silva
Francisco Souza Fadigas

Phytoremediation of soils contaminated by lead and cadmium

Utilisation of castor bean genotypes

ScienciaScripts

Imprint

Any brand names and product names mentioned in this book are subject to trademark, brand or patent protection and are trademarks or registered trademarks of their respective holders. The use of brand names, product names, common names, trade names, product descriptions etc. even without a particular marking in this work is in no way to be construed to mean that such names may be regarded as unrestricted in respect of trademark and brand protection legislation and could thus be used by anyone.

Cover image: www.ingimage.com

This book is a translation from the original published under ISBN 978-3-330-76535-1.

Publisher:
Sciencia Scripts
is a trademark of
Dodo Books Indian Ocean Ltd. and OmniScriptum S.R.L publishing group

120 High Road, East Finchley, London, N2 9ED, United Kingdom
Str. Armeneasca 28/1, office 1, Chisinau MD-2012, Republic of Moldova, Europe
Managing Directors: Ieva Konstantinova, Victoria Ursu
info@omniscriptum.com

ISBN: 978-620-8-51949-0

The first author dedicates the book to his wife Genna, who accompanied him unconditionally throughout the development of the work.

It's easy to be grateful when you see life as a journey of constant learning Unknown

INTRODUCTION

Every day, man's actions leave the world in a situation of increasingly generalised environmental crisis. Natural resources are being exploited uncontrollably, despite the strict environmental legislation implemented in both developed and developing countries. A solution to this problem is still a long way off, because in the world conventions for the preservation of biodiversity and the environment, important G8 countries are not committed to signing protocols, crushing the actions of the so-called G20 countries and others still catching up with development (ROSSET and AVILA, 2008).

Many Brazilian ecosystems have already been compromised by the inadequate exploitation of natural resources and the waste left behind by humanity, contaminating the soil, water and air. Much of this waste is returned to humans through the consumption of contaminated water, plants and animals (PELICIONI, 2005).

Soil is usually contaminated by the decantation of waste produced by industrial and domestic effluents, generally releasing chromium, cadmium, arsenic, barium, mercury and lead. Among these heavy metals, lead has received the most scientific investigation due to the high levels of contamination found in some Brazilian states (LIMA et al. 2010).

It is possible for soil to contain lead naturally, but in quantities that do not pose a danger of contamination to animals and plants. However, human activities in mineral processing industries that store lead, or even oil processing industries, cause the levels of this heavy metal to rise significantly in the soil. It is in the soil that these residues are usually deposited and once there, they remain insoluble, solubilised or can aggregate in organic compounds in the form of colloids, and their mobility is influenced by the pH of the soil itself (BORGES and GOETZ, 2008).

Although it is considered one of the most dangerous heavy metals for human and animal health, lead can be accumulated by plants and does not play a role in their metabolism. When present in the soil, lead is absorbed by plants and they can suffer the process of cell growth inhibition, damaging the development of biomass

(BERTOLI et al. 2011). However, according to Abreu et al. (2009) in soils with high lead concentrations, no phytotoxic effects were observed in tests using concentrations of up to 200 mg/L of soluble lead added to the soil. According to the same authors, most of the lead contained in the soil absorbed by the plant remains in the first 15 cm of depth, and only a small amount can leach out slowly.

Plants do not transport lead in considerable quantities to their aerial parts. Lead can rarely be found in quantities of up to 10 mg/L (BERTOLI et al. 2011). On the other hand, the root system can contain high concentrations, up to 100 per cent more than those found in the upper parts. According to Duarte and Pascoal (2000), the transport of lead from the roots to the aerial part represents only 3%, because unlike the root part, the other organs do not have the same sensitivity to accumulate heavy metals, making the root the main storage organ.

According to Clemens (2001), certain plant species and strains are capable of accumulating large quantities of heavy metals in their tissues and organs. Based on this characteristic, the concept of phytoremediation was developed, where plants are used to clean up contaminated soil and water. The author's work focused on identifying and characterising plant genes that are tolerant to the absorption and accumulation of metals. The same author states that phytochelatins are linear tripeptides that are part of the plant detoxification system. Genes that encode phytochelatin synthesis have been cloned and are now being studied in relation to their regulation, biochemical and biotechnological potential.

Phytoremediation is one of the bioremediation techniques that uses plants to absorb components from contaminated soils (CUNNINGHAM and BERTI, 1993). According to Andrade et al. (2007), this technique is highly versatile and can be used to remediate soil, water or air. According to these authors, phytoremediation has five mechanisms, namely: phytoextraction, where the heavy metal contained in the soil is absorbed and after this mechanism, it is stored in the plant tissue; phytodegradation, where the contaminating element undergoes bioconversion inside or on the surface of the plant, becoming less toxic; phytovolatilisation, where the absorbed contaminant is converted into a volatile substance and released into the atmosphere; phytostimulation, where the presence of plants stimulates microbial biodegradation by components of the root system; and finally phytostabilisation, where the contaminating element is immobilised through lignification or humification.

Several studies have been carried out with plants of agronomic interest for soil phytoremediation. These include sunflower (Alves, 2007), castor bean (Jorge et al. 2010), pork bean (Almeida et al. 2008), forage peanut (Souza, 2010), giant hen's-foot grass (Procópio et al. 2008), rice (Coutinho and Barbosa, 2007), algaroba (Holzbach et al. 2012), vetiver, jureminha and algaroba (Alves et al. 2008), among others.

A very serious case of environmental contamination by heavy metal waste occurred in the Recôncavo region of Bahia. For many years, Companhia Brasileira de Chumbo - COBRAC and later Usina Plumbum Mineração e Metalurgia LTDA operated in the city of Santo Amaro - Bahia, where they processed minerals into lead alloys used in the manufacture of television tubes and batteries. It is estimated that the companies produced more than 850,000 tonnes of lead alloy. This left a liability of more than 500,000 tonnes of waste (cadmium and lead slag), with Cd being found in greater quantities (ALCÂNTARA, 2010). Twenty years ago, in 1993, the activities of this last company were paralysed, without the waste receiving any kind of treatment or proper storage. The slag was used by the public authorities to build landfills, pave streets and avenues in the city and much of this waste was taken and used as fertiliser by small farmers living around the plant's facilities, resulting in contamination of the soil and consequently of the food produced in the affected areas (ANJOS, 2001).

There are some studies in the literature that have tested plants as environmental phytoremediators, such as the study carried out by Alves et al. (2008) to assess lead absorption using vetiver (Vetiveria zizanioides L.), jureminha (Desmanthus virgatus L.) and algaroba (Prosopis juliflora L.).), significant reductions were observed in the dry matter of the root and upper part of the three species as lead concentrations were increased to critical toxicity doses; although among the three species, vetiver was found to be more tolerant of contamination. In vetiver and also in algaroba, there was a greater sensitivity of the root to contamination. The amount of Pb in the internal parts of the plants was also altered by the increase in dosages.

Gonçalves Júnior et al. (2000) carried out an experiment to assess the phytoavailability of cadmium, lead and chromium in soya beans grown on dark red latosol treated with commercial fertilisers. They concluded that the plant showed good bioavailability of these elements in the evaluations. As fertiliser dosages increased, there was an increase in metal displacement in the plant. They also observed that soils with a high amount of organic matter and a high concentration of clay provided greater adsorption of lead and did not make the metal available to the plant. The addition of fertilisers containing heavy metals, the retention of the elements by the pot used in the experiment and the low CEC of the soil increased the availability and displacement of the metals in the plant tissues.

The authors Borges and Goetz (2008) evaluated an area contaminated by Pb after nine years of contact between lead and soil on a farm located in Minas Gerais. For the analysis, they used various local native plants and other edible plants, as well as soil samples. They concluded that even though a long period had passed between the contamination and the analyses, high concentrations of lead were found in the soil, which demonstrated its dangerousness and durability. The same authors suggested that soil removal techniques, even though they have proven to be efficient, have a high financial cost and can generate other environmental impacts. They pointed out that phytoremediation is possible, since bioavailable lead was detected in cover crops, especially in more acidic soil samples.

Duarte and Pasqual (2000) recommended that environmental screening and monitoring using bioindicators should be adopted as fundamental tools, allowing preventative measures to be applied. The quantity of metals found in soil, plants and human hair studied by them is a good indication of the association between environmental pollution and the risk of exposure to humans.

Using sunflower as a phytoremediation plant, Alves (2007) states that these plants accumulate more lead in the roots than in the branches and leaves and that the absorption of higher doses of Pb causes phytotoxicity and consequently the death of the plant.

Nogueira et al. (2008) tested the concentration of lead and other heavy metals in maize plants subjected to soil treated with nine annual applications of sewage sludge. They concluded that the high levels caused by the cumulative effect of chromium, lead and zinc in the roots and aerial parts were increased by successive applications of sewage sludge to the soil. However, they noted the greatest accumulation in the roots. As for the grains, they were well below the levels permitted for human consumption, according to Brazilian legislation.

In another study, it was concluded that the plant species Canavalia ensiformis L., known as the pork bean, has phytoremediation potential for lead, since its growth is not inhibited in the presence of this heavy metal and no symptoms of phytotoxicity occur in the aerial part (ALMEIDA, et al. 2008).

According to Bourlegat et al. (2007), high concentrations of Pb affect the initial development of leucaena plants (Leucena leucocephala Wit.), but do not alter the action of the rhizobia. The rhizobium strain used proved to be resistant to the concentrations of lead applied. The same authors state that inoculated plants grow better in the presence of Pb than non-inoculated plants.

Andrade et al. (2008) stated that there was a greater accumulation of zinc and lead in the roots of rice (contaminated with steel waste), while cadmium was concentrated in the aerial part. The Pb content found in all the samples analysed was well above the average considered phytotoxic.

Hall (2002) reports that heavy metals such as copper and zinc are essential for normal plant growth, although high concentrations imply growth inhibition and the appearance of toxicity symptoms. Plants have a range of potential cellular mechanisms that are involved in detoxification and tolerance to heavy metal-related stress. These include: reduced uptake or flow of metals in the plasma membrane; chelation of metals in the cytosol by peptides such as phytochelatins; repair of proteins damaged by stress.

Experiments carried out by Cruvinel (2009) in Guaxupé/MG evaluated phytoremediation in soil contaminated with the metals nickel, lead, cadmium, chromium and zinc, using three plant species (Brassica juncea, Brachiaria decumbens and Pfaffia glomerata) to recover soil contaminated by these metals. He realised that Brassica juncea was the species that generally performed best in terms of mitigating the contamination of the soil studied, and that it thrived in the different types of contamination with all these metals.

Lima et al. (2010) concluded that the retention of lead in the tissues of the root system, stem and leaves of the castor bean (Ricinus communis L.) indicated that this plant was a good phytoremediator of lead, suggesting its use in effluent polishing systems.

Figueroa et al. (2007) determined the levels of phytochelatins in wild plants, including castor bean, emphasising the relationship between heavy metals and humic acids present in the soil at various sites in the city of Guanajuato in Mexico, which was a centre for silver and gold mining. They also observed that levels of Cd and Pb in plant roots showed a strong positive correlation with phytochelatins, indicating that these two metals promote induction in R. communis. The correlation was inverse between humic substances and metal levels in their roots. Suggesting that metals strongly bound to humus could be less bioavailable to plants, which in turn would limit their role in the induction of phytochelatins.

In view of the reports cited above, it is possible to use plants for phytoremediation of contaminated soils. To this end, Ovsiany and Delai (2007) carried out physiological tests which proved that castor bean plants can be used as phytoremediators, using the phytoextraction process. In this sense, the use of R. communis strains would be a viable auxiliary alternative for use in cleaning up the soil in Santo Amaro, BA.

Another advantage of using castor beans is the production of castor oil, the result of pressing the seeds, which contains 90% ricinoleic fatty acid, giving it special characteristics. This allows it to be widely used industrially, giving the castor bean crop great economic potential for Brazil, making it an alternative source of renewable energy since there are no studies that indicate that the oil is contaminated with heavy metal residues (AMORIM NETO et al. 2001).

Today, castor oil is used to make biodiesel, enamels, resins, lubricants and its biopolymers can be used to build human prostheses. Its stem is used in the production of cellulose and the manufacture of rustic fabrics. As a by-product, the cake has the potential to restore soils with low fertility through cultivation. Its leaves are used as food for silkworms and cows, increasing milk production. Known as green oil, it is compulsorily included in 4% of diesel oil for buses and lorries. It generates jobs in the less favoured regions of the country, involving family farming (CANGEMI et al. 2010).

According to (Freire, 2001) the oils and fatty bodies derived from castor beans result in various industrial applications such as: plastics, soaps, cosmetics, textile dyes, rubbers, hygiene and cleaning agents, toothpaste, pharmaceutical products, edibles, lacquers, synthetic resins, explosives, cellulose processing, mining, biocides, additives for mineral oils, dyes and varnishes.

In this sense, the study of the evaluation of castor bean strains to establish their potential for absorbing lead and cadmium is a very promising alternative for future soil decontamination in rural properties and in the city of Santo Amaro in Bahia. This motivated the development of this study, the aim of which is to evaluate different castor bean strains in the phytoremediation of soils contaminated by lead and cadmium in Santo Amaro - Bahia.

BIBLIOGRAPHICAL REFERENCES

ABREU, M. M., SANTOS, E. S., ANJOS, C. Absorption capacity of lead by spontaneous plants of the genus Cistus in mining environments. Rev. de Ciências Agrárias. Lisbon. v. 32, n. 1. p. 170-181, jan. 2009.

ALCÂNTARA, M. M. Cidade de Chumbo: uma experiência de divulgação em vídeo sobre a contaminação ambiental na cidade de Santo Amaro da Purificação. Diálogos e Ciência, ano IV, n. 12, p. 107-118, mar. 2010.

ALMEIDA, E. L.; MARCOS, F. C. C.; SCHIAVINATO, M. A.; LAGÔA, A. M. M. A.; ABREU, M. F. Growth of Pigeonpea in the Presence of Lead. Bragantia, Campinas, v. 67, n. 3, p. 569-576, 2008.

ALVES, J. C. Evaluation of Sunflower, Castor Bean, Buckwheat and Vetiver as Lead Phytoaccumulators. 2007. 58 f. Dissertation (Master's in Soil and Water Management) - Federal University of Paraíba, Areia. 2007.

ALVES, J. C.; SOUZA, A. P.; PÔRTO, M. L.; ARRUDA, J. A.; TOMPSON JUNIOR, U. A.; SILVA, G. B.; ARAÚJO, R. C.; SANTOS, D. Absorption and distribution of lead in vetiver, jureminha and algaroba plants. Revista Brasileira de Ciência do Solo. n. 32, p 1329-1336, 2008.

AMORIM NETO, M. S.; ARAÚJO, A. E.; BELTRÃO, N. E. M. Climate and soil. In: AZEVEDO, D. M. P.; LIMA, E. F. O agonegócio da mamona no Brasil. 2. ed. Brasília: Embrapa Informação Tecnológica, 2001. cap. 3, p. 63-76.

ANDRADE, A. F. M.; AMARAL SOBRINHO, N. M. B.; MAGALHÃES, M. O. L.; NASCIMENTO, V. S.; MAZUR, N. Zinc, lead and cadmium in rice plants

(Oryza Sativa L.) grown in soil after addition of steel residue. Ciência Rural, Santa Maria, v. 38, n. 7, p. 1877-1885, Oct. 2008.

ANDRADE, J. C. M.; TAVARES, S. R. L.; MAHLER, C. F. O uso de plantas na melhoria da qualidade ambiental. São Paulo: Oficina de textos. 2007. 176 p.

ANJOS, J. A. S. A. Cobrac, Plumbum, Trevisan - Study of environmental liabilities. In: Seminar on heavy metal contamination in Santo Amaro da Purificação, Santo Amaro - Bahia. In: Bahia Análise & Dados, Salvador, v.2, 2001.

BERTOLI, A. C.; CARVALHO, R.; CANNATA, M. G.; BASTOS, A. R. R.; AUGUSTO, A. dos S. Lead toxicity on the content and translocation of nutrients in tomato plants. Biotemas, v. 24, n. 4, p. 7-15, dec. 2011.

BORGES, M. C. R.; GOETZ, C. M. Evaluation of an area contaminated by lead at

Fazenda Boa Esperança. In: 48th Brazilian Chemistry Congress. Rio de Janeiro, 2008. Available at: <http://www.abc.org.br/cbq/2008/trabalhos>. Accessed on 11/06/2013.

BOULEGAT, J. M. G.; ROSSI, S. C.; CHINO, C. E.; SCHIVINATO, M. A.; LAGÔA, A. M. M. A. Tolerance of Leucena leucocephala (Lam.) de Wit to the heavy metal lead. Revista Brasileira de Biociências, Porto Alegre, v. 5, supl. 2, p. 1017-1019, jul. 2007.

CANGEMI, J. M.; SANTOS, A. M.; CLARO NETO, S. The Castor Oil Green Revolution. Química Nova na Escola. v. 32, n. 1, p. 3-8, feb. 2010.

CLEMENS, S. Developing Tools for Phytoremediation: Towards a Molecular Understanding of Plant Metal Tolerance and Accumulation. International Journal of Occupational Medicine and Environmental Health. v. 14, n. 3, p. 235-239, 2001.

COUTINHO, H. D.; BARBOSA, A. R. Phytoremediation: general considerations and utilisation characteristics. Silva Lusitana, v. 15, n. 1, p. 103-117, 2007.

CRUVINEL, D. F. C. 2009. Evaluation of phytoremediation in soils contaminated with metals. 79 f. Dissertation (Master's Degree in Environmental Technology) - Centre for Exact, Natural and Technological Sciences, University of Ribeirão Preto. São Paulo.

CUNNINGHAM, S., BERTI, W. R. The remediation of contaminated soils with green plants; an overview. In vitro cell and plant development biology, v. 29, p. 207-212, 1993.

DUARTE, R. S.; PASQUAL, A. Evaluation of cadmium (Cd), lead (Pb), nickel (Ni) and zinc (Zn) in soils, plants and human hair. Energia na Agricultura. v. 15, n. 1, p. 47-58, 2000.

FIGUEROA, J. A. L.; WROBEL, K.; AFTN, S.; CARUSO, J. A.; CORONA, J. F. G.; WROBEL, K. Effect of some heavy metals and soil humic substances on the phytochelatin production in wild plants from silver mine areas of Guanajuato, Mexico. Elsevier. v. 70, n. 11, p. 2084-2091, Feb. 2007.

FREIRE, R. M. M. Ricinoquímica. In: AZEVEDO, D. M. P.; LIMA, E. F. O agonegócio da mamona no Brasil. Brasília, DF: Embrapa Informação Tecnológica, p. 295-335, 2001.

GONÇAVES JUNIOR, A. C.; LUCHESE, E. B.; LENZI, E. Evaluation of the phytoavailability of cadmium, lead and chromium in soya grown on dark red latosol treated with commercial fertilisers. Química Nova, v. 23, n. 2, p. 173-177, 2000.

HALL, J. L. Cellular mechanisms for heavy metal detoxification and tolerance. The Journal of Experimental Botany, Southampton, v. 53, n. 366, p. 1-11, jan. 2002.

HOLZBACH, J. C.; BARROS, E. I. T. M.; KRAUSER, M. O.; LEAL, P. V. B. Lead: an introduction to extraction and phytoremediation. Journal of Biotechnology and Biodiversity, v. 3, n. 4, p. 178-183, nov. 2012.

JORGE, D. V. C.; RISSATO, S. R.; GALHIANE, M. S.; ALMEIDA, M. V.; RIBEIRO, R.; CECHIN. I. Evaluation of the phytoremediation process of organochlorine pesticides in castor bean and soya samples using different extraction methodologies. In: 50th BRAZILIAN CONGRESS OF CHEMISTRY. 2010. Cuiabá. Proceedings...

LIMA, A. M.; MELO, J. L. S.; MELO, H. N. S.; CARVALHO, F. G. Evaluation of the phytoremediation potential of castor bean (Ricinus communis L.) using synthetic effluent containing lead. Holos, Natal, ano 26, v. 1. p. 51-61, 2010.

NOGUEIRA, T. A. R.; OLIVEIRA , L. R.; MELO, W. J.; FONSECA, I. M.; MELO, G. M. P.; MELO, V. P.; MARQUES, M.O. Cadmium, chromium, lead and zinc in maize plants and soil after nine annual applications of sewage sludge. Revista

Brasileira de Ciência do Solo, n. 32, p. 2195-2207, 2008.

OVSIANY, R; DELAI, R. M. Physiological tests to validate castor bean as a phytoremediation plant. 2007. Available at: <http://www.fag.edu.br/tcc/2007/Ciencias_Biologicas_Bacharelado/>. Accessed on 06 June 2013.

PELICIONI, M. C. F.; JR ARLINDO P. Educação Ambiental e Sustentabilidade. São Paulo. Manole. 191 p. 2005.

PROCÓPIO, S. O.; CARMO, M. L.; PIRES, F. R.; FILHO, A. C.; BRAZ, G. B. P.; SILVA, W. F. P.; BARROSO, A. L. L.; SILVA, G. P.; CARMO, E. L. Phytoremediation of soil contaminated with picloram by giant chickweed (Eleusine coracana). Brazilian Journal of Soil Science. Viçosa. n. 32, 2517-2524, 2008.

ROSSET, P.; AVILA, D. R. Causas de la crisis global. Political Ecology. Institute for Food and Development. California. 2008.

SOUZA, A. M. Phytoremediation of metals in laboratory waste. In: I SIMPÓSIO NACIONAL SOBRE TRATAMENTO DE RESÍDUOS DE LABORATÓRIO. 2010, Jequié. Proceedings...

CHAPTER 1

BIOMETRY AND FLOWER BIOLOGY ASPECTS OF MAMON (Ricinus communis L.) LINES IN SOIL CONTAMINATED BY CADMIUM AND LEAD

INTRODUCTION

The castor bean tree (Ricinus communis L.), which may have originated in Africa, is a dicotyledonous shrub belonging to the Euphobiaceae family. Oil is extracted from its seeds and used mainly as a biofuel, biopolymer for building human prostheses, solvent and resin (SAVY FILHO, 2008; CARNEIRO, 2003). As such, it is a crop of great economic importance for the Brazilian Northeast due to its production potential, which according to Oliveira (2003) can reach up to 6 million hectares per year. It is therefore an excellent source of employment and income for family farmers (PIRES, 2004).

Another possible way of using papaya is in the phytoremediation of soils contaminated by heavy metals (Cunningham and Berti, 1993), as plants naturally absorb the chemical elements present in the soil, which are essential for their development (RAVEN, 2007). Macro and micronutrients are absorbed by the root in two ways. The first through the intercellular spaces and the second inside the cells. When they cross the endodermis, the nutrients reach the xylem and phloem and by active transport are transferred to the interior of the xylem and then transported to the aerial parts of the plant and used for its full development. In the same way that they absorb macro and micronutrients, it is common for plants to absorb other elements present in the soil (FERRI, 1985; SALISBURY and ROSS, 1991 MARSCHNER, 1995; TAIZ and ZEIGER, 1998; HOPKINS, 2000; TAIZ and ZEIGER, 2004).

Because they are considered trace elements, the presence of lead and cadmium is common in most soils in small concentrations (ATSDR, 2009; TORRI and LAVADO, 2009). However, due to numerous anthropogenic activities, the percentage of these and other metals in the soil has increased in concentrations that are harmful to human health (MELO et al. 2008; GOLPAL and RIZVI, 2008).

On the other hand, Chen et al. (2007) stated that the presence of high concentrations of lead and cadmium in the soil is not directly related to their mobility or availability by the plant. This will depend on the plant's physiology and species, with some being more susceptible to absorption and others less so. In this sense, tests with different plants are needed to better assess their potential for phytoremediation of soils contaminated by these heavy metals.

Lead becomes unavailable in the soil due to its adsorption to iron and manganese oxides, and organic matter gives it relative stability (MANECKI et al. 2006). Therefore, plants do not generally transport lead in considerable quantities to their upper parts. Rarely, lead can be found in quantities of up to 10 mg/L. However, the root system can contain high concentrations, up to 100 per cent more than those found in the upper parts. According to Duarte and Pascoal (2000), the transport of lead from the roots to the aerial part represents only 3%, because unlike the root part, the other organs do not have the same sensitivity to accumulate heavy metals, making the root the main storage organ. In this way, even some plants with edible aerial parts could be used in contaminated areas.

Some plants have been tested with positive effects on the absorption of heavy metals such as Pb, Cd and others, among which: sunflower (Lima, 2007), castor bean (Jorge et al. 2010), pigeonpea (Almeida et al. 2008); forage peanut (Gratão et al, 2005); soya bean (Carmo et al. 2008); rice (Coutinho and Barbosa, 2007) and algaroba (Holzbach et al. 2012); vetiver (Alves et al. 2008); leucine (Bourlegat et al. 2007); inedible crucifers (Cruvinel, 2009); guandu beans (Pires et al, 2006); medicinal plants and algae (BAKER, et al. 1994).

In order to test the phytoremediation potential of R. communis L, the aim of this study was to evaluate aspects of the biometry and floral biology of castor bean (Ricinus communis L.) strains in soils contaminated by cadmium and lead.

MATERIAL AND METHODS

Soil analyses were carried out in two different areas, the first (Pb 605 mg/kg and Cd 1.64 mg/kg) taken near the district of Pedra (12°33'H"S 38°46'38"W) approximately 5 kilometres from COBRAC, and the second (Pb 17 000 mg/kg and Cd 92.4 mg/kg) taken 200 metres from the former Companhia Brasileira de Chumbo (COBRAC) (12°33'58"S 38°43'42"W) in Santo Amaro - Bahia. The samples were subjected to chemical analysis (Pb and Cd content) at the SENAI/CETIND-BA Laboratories and physical and chemical analysis at the Embrapa Cassava and Fruit Growing Soil Laboratory, to measure contamination levels and general characteristics of the samples, before the castor bean lines were planted. The soil sample collection points were chosen due to their proximity to the source of contaminant distribution (sample B) and the greater distance from the factory premises where there was supposed to be no contamination (sample A). During the collection and transport of the contaminated soil from the municipality of Santo Amaro to the UFRB in Cruz das Almas, all the people involved wore the appropriate personal protective equipment in order to have as little contact as possible with the metals investigated in this experiment, such as boots, gloves, masks and thick clothing.

The experiment involved sowing in 23kg plastic pots using 5 castor bean strains, namely: L1 - (Lineage 26); L2 - (Lineage 75); L3 - (Lineage 98); L4 - (Lineage 175) and L5 - (Lineage 254) in 4 treatments T1 - (less contaminated soil); T2 - (contaminated soil); T3 - (contaminated soil + organic matter); T4 - (contaminated soil + EDTA chelator) in 4 statistical replications under a completely randomised design. The amount of organic matter used in T3 was 132 grams per pot, corresponding to 13 tonnes per hectare. In T4, the amount of EDTA was 0.19 mol L^{-1} per pot.

In the greenhouse, the plants were kept under natural temperature and photoperiodic conditions. The plants were irrigated every two days, considering 80% of the maximum retention capacity (saturation) of 23.7 g of water per pot, during the 4-month experimental period (Figure 1).

Figure 1 - Castor bean strains subjected to greenhouse conditions in Cruz das Almas, Bahia.

Five seeds were placed in each pot and soon after germination, the seedlings with the least growth were thinned out, leaving only one plant per pot. The pots in the greenhouse were changed every 15 days so that all the plants were exposed to the same conditions.

Four months after planting, biometric assessments were carried out, as shown in Figure 2, such as: stem diameter (SD), measured at 10 cm from ground level; plant height (ST); leaf area (LA) of leaves 1, 2, 3 and 4, using a digital caliper and ruler; number of internodes (INT); number of fruits (NF) and number of seeds (NS). In the latter two, only the strains that showed fruit set.

Studies related to floral biology included analysing flower anthesis, stigma receptivity (Dafni et al. 2005), secretion of extra-floral glands (Galetto and Bernardelo 2004) and viability by colourimetry and pollen germinability by in vitro tissue culture (Ribeiro 2006; Dafni et al. 2005 and Almeida et al. 2004). The data was analysed by variance and the means were compared using the Scott-Knott test at 5% probability.

Figure 2 - Biometric measurements of castor bean lines. A: plant height; B: leaf length 2 and C: stem diameter.

RESULTS AND DISCUSSION

The soil analyses carried out in Santo Amaro - Bahia, in two different areas, allowed the soils to be classified as less contaminated and very contaminated, the results of which are described in Tables 1 and 2. Sample A, collected approximately 5 kilometres from the abandoned factory (COBRAC), was initially designated as uncontaminated soil due to its distance from the factory. However, after analysing the soil in this area, contamination was detected, albeit in smaller quantities of cadmium and lead, in the proportion of 1.64 and 605 mg/kg, respectively (Table 1).

In the second area (sample B), collected in the vicinity of the old factory, the soil was designated as "very contaminated" at 92.4 and 17,000 mg/kg of cadmium and lead, respectively (Table 2).

The "least contaminated" soil was classified as sandy loam and the "most contaminated" soil as sandy loam according to the methodology developed by Embrapa as per the laboratory report (Appendix 1).

According to the laboratory report used by SENAI/SETIND, using the flame atomic absorption spectrometry method, the amounts of Cd and Pb exceed the acceptable limits of 1.6 and 72 mg/kg, respectively, analysed to comply with

CONAMA 420/2009 (Annex 2). These results justify this work given the need for alternatives for soil treatment in the municipality of Santo Amaro - Bahia.

Table 1 - Transcription of the data resulting from the analyses carried out by SENAI/CETIND on the soil samples collected near the District of Pedra, 5 km from COBRAC in Santo Amaro - Bahia.

Sample A: Soil collected 5 kilometres from the abandoned factory			
Essay	**Heavy metal content mg/kg**	**Acceptable limit (L1) ***	**LQ****
Cd	**1,64**	1.3 mg/kg	0,49
Pb	**605**	72 mg/kg	4,5

*CONAMA 420/2009 - contaminated soil

** Limit of Quantification

Table 2 - Transcription of the data resulting from the analyses carried out by SENAI/CETIND with a soil sample collected 200 m from COBRAC in Santo Amaro - Bahia.

Sample B: Soil collected in the vicinity of the abandoned factory			
Essay	**Heavy metal content mg/kg**	**Acceptable limit (L1) ***	**LQ****
Cd	**92,4**	1.3 mg/kg	0,49
Pb	**17000**	72 mg/kg	4,5

*CONAMA 420/2009 - contaminated soil

** Limit of Quantification

It can be seen from the data in Tables 1 and 2 that there is contamination well above the limits acceptable under environmental legislation, which could jeopardise the development of local agriculture and, therefore, the development of the environment.

development of the metabolism of living beings in general. Machado et al. (2013) reported on the risks of contaminated soil in Santo Amaro, whether through the introduction of this heavy metal into the food chain or through the aspiration of soil dust and even geophagy, comprising significant risks for the population.

A study carried out by Tavares (1992) stated that metals such as Pb and Cd are considered toxic to humans and are made available in the environment due to their widespread industrial use and, consequently, from a toxicological point of view. These elements react with macromolecules present in cell membranes, giving them bioaccumulative and spreading properties through the various levels of trophic chains, transforming concentrations considered normal into toxic for various species, particularly humans. Metals are bioaccumulative, persisting for long periods even after emissions have ceased. Naturally, metals that do not fulfil biological functions are found in concentrations in the part per billion (ppb) or part per trillion (ppt) range. Increasing values of these elements go from tolerable to toxic. In this sense, the values of these heavy metals found in the samples analysed exceed the tolerated values.

The study of flower opening (anthesis) is fundamental for observing the normality of the full development of plant species (Almeida et al. 2004; Dafni et al. 2005; Raven, 2007), as this event guarantees the continuity of reproduction, especially under the conditions of this study, where the strains under study were subjected to different planting treatments.

In all the treatments with strain 1 (L1) there was no flowering, which could characterise this strain as not very tolerant to absorbing the heavy metals under study. However, further studies should be carried out with this strain. On the other hand, anthesis in the treatments with the strains (L2T1 - less contaminated soil; L5T1 - less contaminated soil; L2T2 - contaminated soil; L3T2; L4T2; L5T2 - contaminated soil; L2T3 - contaminated soil + organic matter; L3T3 - contaminated soil + organic matter; L4T3 - contaminated soil + organic matter; L5T3 - contaminated soil + organic matter; L2T4 - contaminated soil + chelating agent; L3T4 - contaminated soil + chelating agent; L4T4 - contaminated soil + chelating agent and L5T4 - contaminated soil + chelating agent) occurred from 5:00 h with the beginning of the opening of the female and male flowers synchronously with a durability of 24 hours of the female flower and up to 48 hours of pollen dispersal in the male flowers (Figure 3). However, in the treatments with the strains (L1T1; L3T1; L4T1; L1T2; L1T3 and L1T4) there was no flowering during this period.

Weiss (2000) reported on the complexity of the physiology of the papaya tree, which has a C3 metabolism. Its panicle inflorescence has male flowers in the lower portion and female flowers in the upper portion (SILVA, 1999; SILVA et al. 2007; FONSECA et al. 2012). These studies corroborate the floral specificities found in the present study.

Figure 3 - Stages of castor bean flower development: A (Initial inflorescence formation (L3 - Lineage 98); B (Synchronisation of male and female flower opening (L2 - Lineage 75); C (Flowers in the female stage (L3 - Lineage 98); D (Flowers in the male stage (L4 - Lineage 176) and (Flowers in the male stage dispersing pollen within 48 hours of anthesis (L5 - Lineage 254).

According to Dafni et al. (2005), stigma receptivity is characterised by the possibility of pollen grains adhering to the stigmatic papillae. In this sense, the stigma is receptive when there is a release of oxygen bubbles under its surface resulting from the breakdown of peroxide by the action of the enzyme peroxidase. In this study, the stigma receptivity test was positive for all the strains where flowering occurred from the opening of the female flower until senescence, with the enzyme peroxidase acting when the stigmatic papillae were subjected to the Hydrogen Peroxide test (Figure 4). This means that the flowers of these strains were normal in terms of pollen grain receptivity in the process of self-fertilisation or dispersal by some pollinating agent, even when the plants were subjected to contaminated soil conditions, corroborating the data of DAFNI et al. (2005); DAFNI, et al. (1992); SILBERBAUER-GOTTSBERGER, (1998) and RICHARDS (1997).

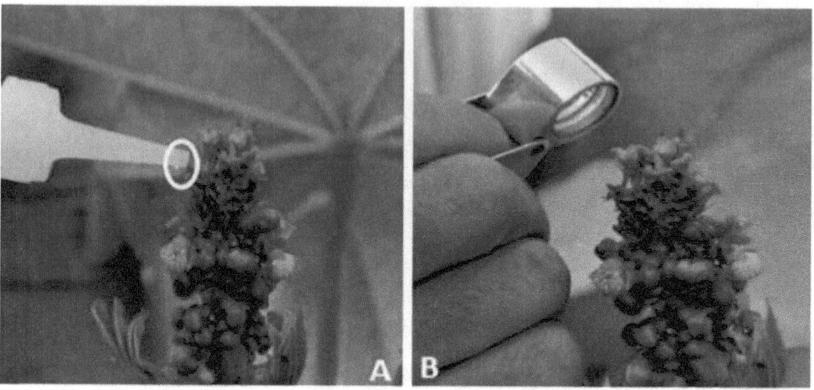

Figura 4 - Stigma receptivity test. A: detail of the formation of oxygen bubbles due to the action of peroxidase and B: observation of bubble formation using a hand-held magnifying glass.

No nectar secretion was observed in the floral nectary chamber. However, it was observed in the flowering lines that there is a sweet secretion with a Brix equal to 25% released by the extrafloral nectaries (Figure 5). This secretion may explain the constant visitation by bees, wasps and ants on castor bean lines in field conditions, and it is also the reason why the plant is constantly being visited by bees, wasps and ants.

according to some authors. Baker et al. (1978), for example, reported that castor bean trees have six to seven extrafloral nectaries and that their flowers are polliniferous. Milfont et al. (2009) observed the collection of extrafloral nectar by Apis mellifera bees in a castor bean monoculture area. Taiz and Zeiger (2004) reported that extrafloral nectaries function as drains of the plant's reserves and nutrients. Dafni et al. (2005) and Galetto and Bernardelo (2004) stated that the extrafloral nectaries can also serve as protection for the plant, since these structures attract many ants and these keep natural enemies away from the plant.

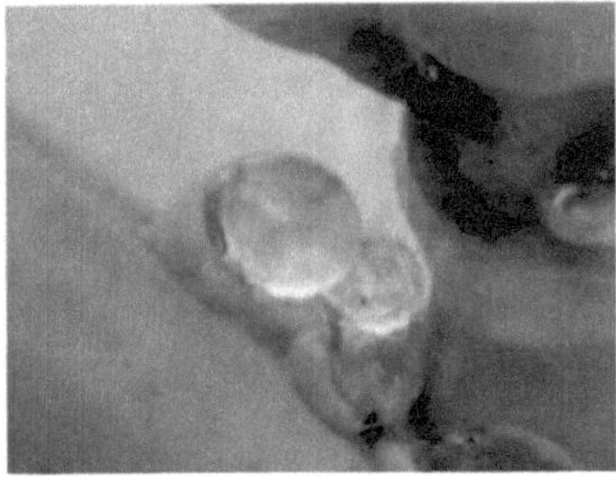

Figura 5   - Presence of sweet secretion in the extrafloral nectaries with Brix of 24% in castor bean strains.

The results of the pollen viability and germination rates of the strains tested showed a highly significant interaction between the parameters tested and the strains (p< 0.05) using the Scott-Knott test at the 5% probability level, demonstrating different behaviour as a result of the variations in pollen evaluation periods. The high pollen viability rates may be related to the high degree of viability by colourimetry when using some dyes, including acetic carmine. Shivanna and Johri (1992) stated that it is possible for pollen viability estimates to vary between plant species or even between individuals of the same species and different lineages.The in vitro pollen germinability technique makes it possible to check for contrasts in viability and germinability data. In this test, there was also a significant difference between the treatments (Annex 3 and Figures 6, 7 and 8).Loguercio and Battistin (2004) stated that pollen viability tests determined using dyes that act on the pollen's genetic material can overestimate the percentage of male fertility, while in vitro or in situ germinability tests can underestimate this percentage. On the other hand, Kearns and Inouye (1993) considered colourimetry to be an indirect method and in vitro germination to be a direct method, both of which are satisfactory for determining the genetic viability of plants. Justo et al. (2008) found pollen viability and germination rates similar to those found in this study. The authors used acetic carmine to test viability and basic culture medium for germinability. They concluded that the colour (viability) of the pollen grains does not always overestimate the germinability percentage, as they obtained similar averages. The mean viability and pollen germination data from this study show that the male fertility of the strains tested was not affected by contamination of the soil in which they were planted.

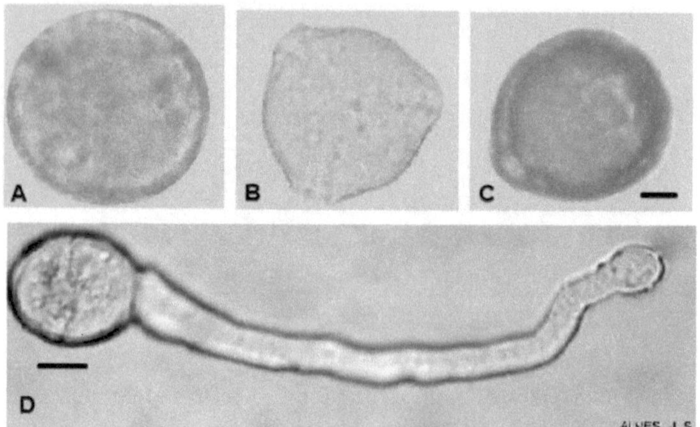

Figura 6 - Standard pollen viability and germinability of five castor bean strains. A: viable pollen grain in polar position I stained with acetic carmine; B: non-viable pollen grain (not stained) in polar position II; C: pollen stained with ammonium blue (germinability test) in equatorial position and D: pollen germinated shortly after anthesis. Scale: 12-14pm.

The radar graphs (Figures 7 and 8) show the distribution of the male fertility rate of the four strains tested as a function of the colourimetry techniques and in vitro pollen germinability ($p < 0.05$), demonstrating their good adaptation. Therefore, they were not affected by the cadmium and lead contaminated soils in Santo Amaro - Bahia.

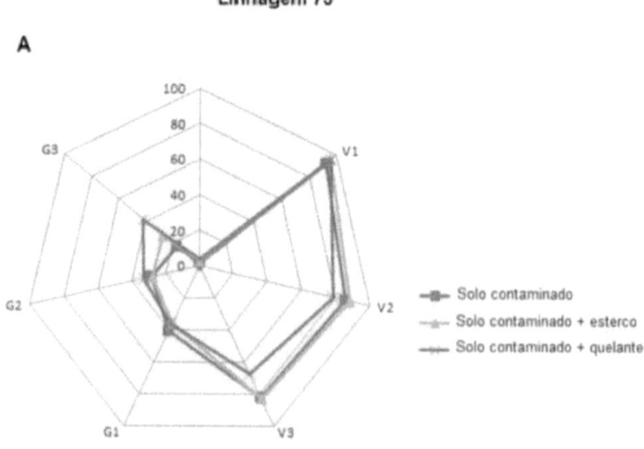

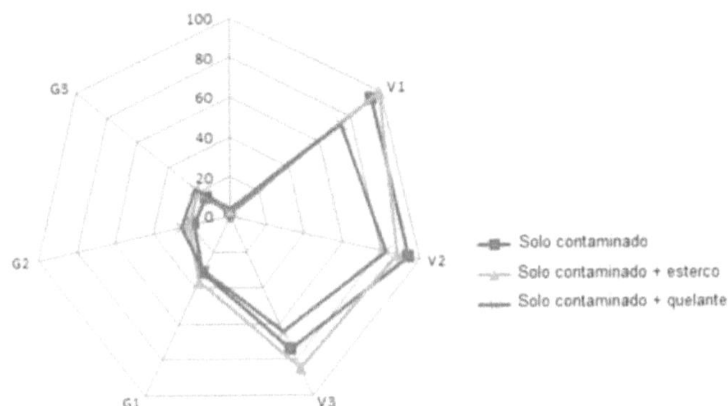

Figura 7 - Estimation of pollen viability and germinability of Strains **A** - L2 (Lineage 75) and **B** - L3 (Lineage 98): V1= pollen viability at anthesis; V2= pollen viability 8 days after anthesis; V3= pollen viability 15 days after anthesis; G1= pollen germinability at anthesis; G2= pollen germinability 8 days after anthesis and G3= pollen germinability 15 days after anthesis.

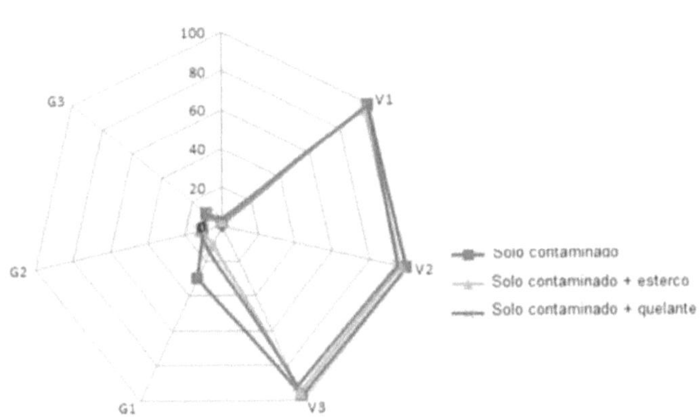

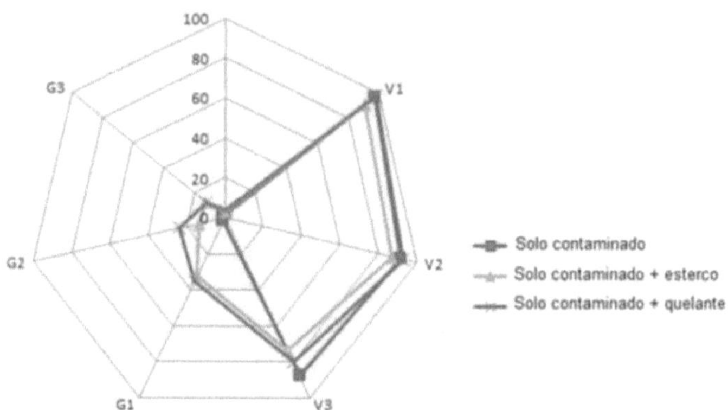

Figura 8 - Estimation of pollen viability and germinability of Lineages **A** - L4 (Lineage 176) and **B** - L5 (Lineage 264): V1= pollen viability at anthesis; V2= pollen viability 8 days after anthesis; V3= pollen viability 15 days after anthesis; G1= pollen germinability at anthesis; G2= pollen germinability 8 days after anthesis and G3= pollen germinability 15 days after anthesis.

When assessing the biometric measurements (Table 4), the analysis of variance showed that there was a significant difference (p< 0.05) between treatments T1 - less contaminated soil; T2 - contaminated soil; T3 - contaminated soil + manure and T4 - contaminated soil + chelating agent. In the comparisons of means (Table 5), it can be seen that treatment 3 (contaminated soil + manure) showed the best performance in most of the characters assessed, differing from the other treatments. Treatments 2 and 4 were similar and did not differ according to the Scott-Knott test (p<0.05) for all the characters assessed. In the work carried out by Nakagawa (1976), an increase in the production of grains in the first bunch was observed in papaya trees submitted to treatment with sewage sludge + fertiliser.

On the other hand, Nascimento et al. (2011) carried out a study on the growth and productivity of castor bean seeds treated with sewage sludge, testing 5 doses on a dry basis (0; 15; 30; 45 and 60 t ha), observing average values for the characteristics of plant growth and seed productivity. These biometric characteristics increased as the doses increased, with the dose starting at 15 t ha^{-1} being sufficient to replace the crop's mineral fertiliser without the risk of contaminating the soil with heavy metals.

Table 4 - Summary of the Analysis of Variance relating to growth (biometric analysis) in castor bean lines at 120 days of cultivation, when the seeds were planted in soil contaminated with Cadmium (Cd) and Lead (Pb) and subjected to four treatments.

				Mean Square				
FV	**GL**	**EST**	**DC**	**AF1**	**AF2**	**AF3**	**AF4**	**INT**
Treatments	3	617,85	21,98	119152,9	96845,1	85066,83*	53917,16*	50,1*
Waste	16	14,31	1,07	9701,67	5660,9	9336,14	10163,9	7,9
CV(%)	-	7,97	10,61	28,37	22,84	37,71	55,67	14,9

*Highly significant effect by F test at 5% probability level

Captions: EST (plant height) DC (stem diameter); AF1 (leaf area 1); AF2 (leaf area two); AF3 (leaf area three); AF4 (leaf area 4); INT (internode).

Table 5 - Characters related to growth (biometric analysis) in castor bean lines at 120 days of cultivation, when the seeds were planted in soil contaminated with Cadmium (Cd) and Lead (Pb) and subjected to four treatments.

				Characters			INT
Treatment.	**EST (cm)**	**DC (mm)**	**AF1 (cm)**	**AF2 (cm)**	**AF3 (cm)**	**AF4 (cm)**	
T-1	35,2 c	7,08 c	142,72 c	142,40 c	83,31 c	39,09 b	14,2 b
T-2	44,5 b	9,50 b	329,5 b	324,34 b	247,5 b	170,67 a	19 b
T-3	62 a	12,09 a	506,76 a	473,86 a	394,9 a	272,56 a	21 a
T-4	47,95 b	10,48 b	409,7 b	376,76 b	298,8 b	242,56 a	20,8 a

CONCLUSION

There is soil contamination, even if it is considered low, within a 5-kilometre radius of the abandoned factory with lead slag liabilities;

Male fertility, stigmatic receptivity and the production of extrafloral nectaries were not affected by the treatments of the 4 tested castor bean strains;

The strains evaluated performed better in terms of agronomic traits when submitted to the treatment corresponding to contaminated soil + manure.

BIBLIOGRAPHICAL REFERENCES

ATSDR - AGENCY FOR TOXIC SUBSTANCES AND DISEASE REGISTRY: U. S. DEPARTMENT OF HEALTH AND HUMAN SERVICES. Draft toxicological profile for perfluoroalkyls. 2009. Available at: <http://www.atsdr.cdc.gov/>. Accessed on: 03 May 2013.

ALMEIDA, E. L.; MARCOS, F. C. C.; SCHIAVINATO, M. A.; LAGÔA, A. M. M. A.; ABREU, M. F. Growth of pigeonpea in the presence of lead. Bragantia, Campinas, v. 67, n.3, p. 569-576, 2008.

ALMEIDA, O. S.; SILVA, A. H. B.; SILVA, A. B.; AMARAL, C. L. F. Study of the floral biology and reproductive mechanisms of lavender (Ocimum oficinalis L.) with a view to genetic improvement. Acta Scientiarum Biological Sciences, Maringá, v. 26, n. 3, p. 343 - 348. 2004.

ALVES, J. C.; SOUZA, A. P.; PÔRTO, M. L.; ARRUDA, J. A.; JUNIOR, U. A. T.; SILVA, G. B.; ARAÚJO, R. C.; SANTOS, D. Absorption and distribution of lead in vetiver, jureminha and algaroba plants. Revista Brasileira de Ciência do Solo, Viçosa, v. 32, n. 3, p. 1329-1336. mai./jun. 2008.

ALVES, J.C. 2007. Evaluation of Sunflower, Castor Bean, Buckwheat and Vetiver as Lead Phytoaccumulators. 58 f. Dissertation (Master's in Soil and Water Management) - Federal University of Paraíba, Areia.

BAKER, A. J. M., MC GRATH, S.P., SIDOLI, C. M. D., REEVES, R. D. The Possibility of in situ heavy metal decontamination of polluted soils using crops of metal-accumulating plants. Resources, conservation and Recycling, Amsterdam, v. 11, p. 41-49, jun. 1994.

BAKER, D. A.; HALL, J. L.; THRORP, J. R. Study of the Extrafloral Nectaries of Ricinus communis. New Phytologist, v. 81, n. 1, p. 129-137, 1978.

BOULEGAT, J. M. G.; ROSSI, S. C.; CHINO, C. E.; SCHIVINATO, M. A.; LAGÔA, A. M. M. A.. Tolerance of Leucena leucocephala (Lam.) de Wit to the heavy metal lead. Revista Brasileira de Biociências, Porto Alegre, v. 5, supl. 2, p. 1017-1019, jul. 2007.

CARNEIRO, R. A. F. Biodiesel Production in Bahia. Conjuntura & Planejamento, Salvador, n.112, p. 35-43, Sep. 2003.

PROCÓPIO, S. O.; CARMO, M. L.; PIRES, F. R.; FILHO, A. C.; BRAZ, G. B. P.; SILVA, W. F. P.; BARROSO, A. L. L.; SILVA, G. P.;

CARMO, E. L. Phytoremediation of soil contaminated with picloram by giant chickweed (Eleusine coracana). Brazilian Journal of Soil Science. Viçosa. n. 32, 2517-2524, 2008.

CHEN, S.; SUN, L.; SUN, T.; CHAO, L.; GUO, G. Interaction between cadmium, lead and potassium fertilizer (K_2SO_4) in a soil-plant system. Environ Geochem Health, Amsterdam, v. 29, p. 435-446, Apr. 2007.

COUTINHO, H. D.; BARBOSA, A. R. Phytoremediation: general considerations and utilisation characteristics. Silva Lusitana, Lisboa, v. 15, n. 1, p. 103-117, jun. 2007.

CRUVINEL, D. F. C. Evaluation of phytoremediation in soils subjected to metal contamination. 2009. 79 f. Dissertation (Master's in Environmental Technology) - Centre for Exact, Natural and Technological Sciences, University of Ribeirão Preto. São Paulo.

CUNNINGHAM, S., BERTI, W. R. The remediation of contaminated soils with green plants; an overview. In vitro cell and plant development biology, v. 29, p. 207-212, 1993.

DAFNI, A. Pollination ecology: a practical approach. Oxford: Oxford University Press. 1992. 250 p.

DAFNI, A.; KEVAN, P. G.; HUSBAND, B. C. Practical pollination biology. Cambridge: Enviroquest. 2005. 590 p.

DUARTE, R. S.; PASQUAL, A. Evaluation of cadmium (Cd), lead (Pb), nickel (Ni) and zinc (Zn) in soils, plants and human hair. Energia na Agricultura, v. 15, n. 1, p. 47-58. 2009.

FERRI, M. G. Fisiologia Vegetal. 2. ed. São Paulo: EPU, v. 1, 1985. 361p.

FONSECA, E. R.; EICHOLZ, M. D.; EICHOLZ, E. D.; AIRES, R. F. SILVA, D. A.
Characterisation of the flowering of the castor bean cultivars 'AI guarany' 2002 IAC80. In: 21st CONGRESS OF SCIENTIFIC INITIATION OF THE FEDERAL UNIVERSITY OF PELOTAS, 4., 2012, Pelotas.

GALETTO, L.,; BERNARDELLO, G. Floral nectaries, nectar production dynamics and chemical composition in six Ipomoea species (Convolvulaceae) in relation to pollinators. Annals of Botany, v. 94, n. 2, p. 269-280, 2004.

GOPAL, R.; RIZVI, A, H. Excess lead alters growth, metabolism and translocation of certain nutrients in radish. Chemosphore, Amsterdam, v. 70, p. 1539-1544, feb. 2008.

GRATÃO, P. L.; PRASAD, M. N. V.; CARDOSO, P. F.; LEA, P. J.; AZEVEDO, R. A. Phytiremediation: green technology for the cleanup of toxic metals in the environment. Brazilian Journal of Plant Physiology, Londrina, v. 17, n. 1, p. 53-64, jan./mar. 2005.

HOLZBACH, J. C.; BARROS, E. I. T. M.; KRAUSER, M. O.; LEAL, P. V. B. Lead: an introduction to extraction and phytoremediation. Journal of Biotechnology and Biodiversity, v. 3, n. 4, p. 178-183, nov. 2012.

HOPKINS, W. G. Introduction to Plant Physiology. 2. ed. New York: John Wiley & Sons, Inc., 2000. 512 p.

JORGE, D. V. C.; RISSATO, S. R.; GALHIANE, M. S.; ALMEIDA, M. V.; RIBEIRO, R.; CECHIN. I.. Evaluation of the phytoremediation process of organochlorine pesticides in castor bean and soya samples using different extraction methodologies. In: 50th BRAZILIAN CONGRESS OF CHEMISTRY, 2010, Cuiabá, Anais ....

JUSTO, P. S.; CUCHIARA, C. C.; ANJOS, S. D. LUCIA, V.. Pollen viability analysis of castor bean cultivars (Ricinus communis l.) subjected to gamma radiation. In: III CONGRESSO BASILEIRO DE MAMONA, 2008, Anais...

KEARNS, C. A.; INOUYE, D. W. Techinques for pollination biologists. Niwot: University Press of Colorado, 1993, 579 p.

LIMA, C. V. S.; SILVA, A. A.; SOUZA, E. D.; MEURER, E. J.; ANGHINONI , I; SCHMIDT, R. O. Bioaccumulation of lead by sunflower in arenic red-dystrophic-yellow clay. In: XXXI BRAZILIAN CONGRESS OF SOIL SCIENCE, 2007, Gramado. Proceedings...

LOGUERCIO, A. P.; BATTISTIN, A.. Microsporogenesis of nine accessions of Syzygium cumini (L.) Myrtaceae from Rio Grande do Sul - Brazil. Revista da FZVA, v. 11, n. 1, 2004.

MANECKI, M; BOGUCKA, A; BADJA, T; BORKIEWICZ, O. Decrease of Pb bioavailability in soils by addition of phosphate ions. Environmental Chemistry Letters, v. 3, n. 4, p. 178-181, jan. 2006.

MARSCHNER, H. Mineral Nutrition of Higher Plants. 2. ed. London: Academic Press. 1995. 889 p.

MELO, E. E. C.; NASCIMENTO, C. W. A.; SANTOS, A. C. Q.; SILVA, A. S. Availability and fractionation of Cd, Pb, Cu and Zn as a function of pH and soil incubation time. Ciência Agrotécnica, Lavras, v. 32, n. 3, p. 776-784, mai./jun. 2008.

MILFONT, M. O.; FREITAS, B. M.; RIZZARDO, R.; GUIMARAES, M. O. Honey production by Africanised bees in castor bean plantations. Ciência Rural, Santa Maria, v. 39, n. 4. p. 1206-1211, jul. 2009.

NAKAGAWA, J. Effects of phosphorus on two cultivars of papaya (Ricinus communis L.) Campinas and Guarany. Botucatu, Universidade Estadual Paulista Júlio de Mesquita Filho. 1976. 115p.

NASCIMENTO, A. L.; SAMPAIO, R. A. BRANDÃO JUNIOR, D. S.; JUNIO, G. R. Z.; FERNANDES, L. A. Growth and productivity of castor bean seed treated with sewage sludge. Revista Caatinga, Mossoro, v. 24, n. 4, p. 145-151, Oct-Dec. 2011.

OLIVEIRA, D. Development of castor bean in the northeast is supported by Embrapa. Embrapa Cotton Online. Campina Grande, PB, Jul. 2003. Available at: <http://www.embrapa.br/imprensa/noticias/2003/julho/bn.2004-   11-25.2520470505>. Accessed on 02 May 2013.

PIRES, F. R.; PROCÓPIO, S. O.; SOUZA, C. M.; SANTOS, J. B.; SILVA, G. P. Green manures in the phytoremediation of soils contaminated with the herbicide Tebuthiuron, Caatinga, Mossoró, v. 19, n. 1, p. 92-97, jan./mar. 2006.

PIRES, F. R.; PROCÓPIO, S. O.; SOUZA, C. M.; SANTOS, J. B.; SILVA, G. P. Green manures in the phytoremediation of soils contaminated with the herbicide Tebuthiuron, Caatinga, Mossoró, v. 19, n. 1, p. 92-97, jan./mar. 2006.

RAVEN, P. H.; EVERT, R. F.; EICHHORN, S. E. Biologia Vegetal. Translation by Jane Elizabeth Kraus et al. 7. ed. Rio de Janeiro: Guanabara Coogan, 2007. 830 p.

RIBEIRO, G. S. Aspects of floral biology related to seed and fruit production in the pine cone (Annona squamosa L.). 2006. 72 f. Dissertation (Master's Degree in Agricultural Sciences) - State University of Southwest Bahia, Vitória da Conquista, 2006.

RICHARDS, A. J. Plant Breeding Systems. 2. ed. Chapman. 1997

SALISBURY, F. B., ROSS, C. W. Plant Physiology. 4. ed. California: Wadsworth Publishing Company, Inc., 1991. 682 p.

SAVY FILHO, A.; REGITANO NETO, A.; KIIH, TAMMY A. M. Strategy for the Genetic Improvement of Castor Beans. In: III Brazilian Castor Bean Congress, 2008, Salvador. Proceedings... IAC,

SHIVANNA, K. R.; JOHRI, B. M. The angiosperm pollen: structure and function. New Dehli: Wiley Eastern Ltd. 1992.

SILBERBAUER-GOTTSBERGER, I. Evolution of flower structure and pollination. In Neotropical cassinae (Caesalpiniaceae) species. Phyton, n. 28, p. 293-320. 1998.

SILVA, M. S.; SOUZA, S. N.; LENZI, E.; LUCHESI, E. B. Lead behaviour in clay soil treated with combined sewage sludge and its absorption by plants. Acta Scientiarum, v. 21, p. 757-762, 1999

SILVA, S. D. A. et al. (Ed.). Castor bean cultivation in Rio Grande do Sul. Pelotas: Embrapa Temperate Climate - Production Systems 11. 2007. 115 p.

TAIZ, L., ZEIGER, E. Plant Physiology. 2. ed. Massachusetts: Sinauer Associates, 1998. 792 p.

TAIZ, L., ZEIGER, E. Plant Physiology. Translation by Eliane Romano Santarém et al. 3. ed. Porto Alegre: Artmed, 2004. 719 p.

TAVARES, T.; CARVALHO, M. Evaluation of exposure of human populations to heavy metals in the environment: examples from the Recôncavo baiano. Química nova, v. 15, n. 2, 1992.

TORRI, S.; LAVADO, R. Plant absorption of trace elements in sludge amended soils and correlation with soil chemical speciation. Journal of Hazardous Materials, Amsterdam, v. 166, p. 1459-1465. jul. 2009.

WEISS, E. A. Oilseed crops. London: Blackwell Science, 2000. 364 p.

CHAPTER 2

EVALUATION OF MAMON (Ricinus communis L.) LINES IN THE PHYTOREMEDIATION OF SOIL CONTAMINATED BY CADMIUM AND LEAD IN THE SOILS OF SANTO AMARO, BAHIA[1]

INTRODUCTION

Environmental pollution caused by heavy metals has been increasing in recent decades due to human activities related in most cases to mining, disposal of industrial waste, sewage sludge, fertilisers and agrochemicals in the soil (FREITAS et al. 2009). Trace elements deposited in the soil are adsorbed through ionic exchange, electrostatic interactions, Van der Waals forces or firmer, more stable chemical bonds (LINHARES et al. 2009). When these stable interactions with soil particles do not occur, they migrate, contaminating water and living beings.

Due to its toxic potential and great persistence in the environment, lead is one of the elements that has aroused the most interest among heavy metals, and is considered the second most dangerous element for the environment, according to the US Environmental Protection Agency's list of priorities (ATSDR, 2009; TORRI and LAVADO, 2009). Lead is naturally present in the soil in low concentrations, causing no damage to health. However, anthropogenic intervention can lead to an increase in the content of this element, resulting in the contamination of soils used for growing food (MELO et al. 2008; GOPAL and RIZVI, 2008).

The large amount of lead in the soil is not directly related to its mobility or availability (CHEN 2007). It becomes less available in the soil due to its adsorption to iron and manganese oxides and organic matter, giving it relative stability (MANECKI et al. 2006). The bioavailability of Pb varies according to the characteristics of the soil, the element itself and the length of time the metal remains in the environment (LU et al. 2005).

Another trace element, cadmium (Cd), is usually associated with Pb in contaminated soils and its action directly affects the osmotic balance of plant cells and can also influence the absorption, displacement and assimilation of elements such as Mg, K, P, Ca and Fe by plant structures (DAS et al. 1997; SANITÁ and GABRIELLI, 1999). The Cd-Fe ratio negatively interferes with photosynthetic metabolic processes. In the mechanism for transporting elements between membranes, Cd can replace Ca ions by immobilising it. The absorption of Mn and Mg is impaired by Cd interference (KABATA-PENDIAS and PENDIAS, 2000). According to Guimarães, (2008) Cd, even in low concentrations, can cause protein denaturation, a reduction in the speed of enzymatic and photosynthetic activity and damage to cell membranes as well as chlorosis. Some species show tolerance mechanisms, as is the case with some plants that are phytoextractors of contaminating elements from the soil.

In order to assess the bioavailability of these metals in the soil, some simple extraction techniques using EDTA, DTPA and CaCl2 have been widely used. These techniques can also provide information on the potential for mobility, leaching and soil pollution (ISHIKAWA et al. 2009). Thus, excessive amounts of potentially bioavailable cadmium and lead can be absorbed and accumulated in plants grown in a contaminated environment without showing visible signs of toxicity (PIECHALAK, et al. 2002; HONG et al. 2008).

According to Andrade et al. (2007), phytoremediation is a highly versatile technique that can be used to remediate soils contaminated by heavy metals and other substances. According to these authors, phytoremediation has five mechanisms, namely: phytoextraction, where the heavy metal contained in the soil is absorbed and after this mechanism, it is stored in the plant tissue; phytodegradation, where the contaminating element undergoes bioconversion inside or on the surface of the plant, becoming less toxic; phytovolatilisation, where the absorbed contaminant is converted into a volatile substance, which is released into the atmosphere; phytostimulation, where the presence of plants stimulates microbial biodegradation by components of the root system; and finally phytostabilisation, where the contaminating element undergoes immobilisation through lignification or humification.

The species and physiology of the plant, as well as environmental factors such as pH, soil particle size, cation exchange capacity, organic matter and nutrient availability interfere with the content of Pb and Cd absorbed and accumulated in plant tissues (LU et al. 2005). In plants, most of the Pb absorbed is deposited preferentially in the roots, but it can be transferred and allocated to other parts, including edible parts (LIMA 2010).

The aim of this study was to evaluate five castor bean strains in the phytoremediation of soils contaminated by cadmium and lead in Santo Amaro - Bahia.

MATERIAL AND METHODS

The experiment was carried out in a greenhouse from March 2012 to June 2013 under cover at the Genetic Improvement and Biotechnology Centre (NBIO) located on the campus of the Federal University of Recôncavo da Bahia (UFRB), in the municipality of Cruz das Almas/Bahia. Eighty 23 kg pots (34 cm x 28 cm) were used, containing soil collected from a depth of 0 to 30 cm in the city of Santo Amaro - Bahia, near the former Companhia Brasileira de Chumbo (COBRAC) lead processing plant and within a 5 km radius of COBRAC.

It should be noted that when collecting and transporting the material from the municipality of Santo Amaro to the UFRB in Cruz das Almas, Bahia, all the people involved used individual protection equipment such as gloves, boots, masks and thick clothing in order to minimise contact with the pollutants in question.

Five strains of castor bean (Ricinus communis L.) developed by the UFRB NBIO Genetic Improvement and Biotechnology Centre were used, namely: L1 - (Lineage 26); L2 - (Lineage 75); L3 - (Lineage 98); L4 (Lineage 176) and L5 - (Lineage 254).

The amount of organic matter used in T3 was 132 grams per pot, corresponding to 13 tonnes per hectare. In T4, the amount of EDTA was 0.19 mol L^{-1} per pot.

In the greenhouse, the plants were kept under natural temperature and photoperiodic conditions. The plants were irrigated every two days, considering 80% of the maximum retention capacity (saturation) of 23.7 g of water per pot, during the 4-month experimental period.

The statistical design was entirely randomised with 4 replications using the following treatments: T1: least contaminated soil; T2: soil contaminated with Pb and

Cd slag; T3: soil contaminated with Pb and Cd slag + animal manure (10 t/ha = 102 g per pot); T4: soil contaminated with Pb and Cd slag + EDTA chelator (10 mmol/kg of soil).

After filling the pots, 3 seeds were placed at a depth of 8 and 10 cm in each pot. The first irrigation was carried out according to calculations to determine the maximum water retention capacity (saturation). The positions of the pots were alternated randomly at 15-day intervals.

As a result of the chemical and physical analyses of the soil for micro and macronutrients carried out by Embrapa Cassava and Fruit Growing - Cruz da Almas/Bahia (report attached), fertilisation and soil correction were not necessary. Before the experiment was carried out, Pb and Cd contaminants were quantified in the soil samples by SENAI - CETIND in the municipality of Lauro de Freitas/Bahia in the two soil collection areas. The area near the former COBRAC factory was classified as highly contaminated with high Cd and Pb values. The soil collected in an area within a 5 kilometre radius of the factory was classified as less contaminated. The area selected for the collection of witness material (without contamination) was supposedly free of pollutants, however, after analysing the soil chemically, it was found to contain less cadmium and lead, making it less contaminated.

The experiment was carried out over approximately 120 days, when most of the plants reached the fruiting stage and were stopped when the first signs of senescence appeared. The plants were cut 10 cm from the root with stainless steel pruning pliers. They were washed with drinking water and deionised water and dried under cover from the rain for 72 hours. They were then packed in paper bags, weighed and taken to the 65°C ventilation oven for 48 hours in the Analytical Chemistry laboratory at UFRB. The roots were washed in a fine mesh sieve to avoid loss of material.

The samples were removed from the oven and cooled in a desiccator, then weighed on a semi-analytical balance. They were then ground in a stainless steel knife mill with a 1mm mesh sieve. Careful cleaning was always carried out between samples to avoid contamination. The order of grinding was from the least contaminated treatments to the most contaminated, the first referring to sample A followed by sample B. The material from each plant was placed in sterile plastic jars.

To quantify the levels of Cd and Pb in plant tissues, the samples were digested by the acid digestion method in a digestion block using cold fingers at the Analytical Chemistry Laboratory of the Chemistry Institute of the Federal University of Bahia (UFBA). To do this, 0.2 g of sample and 4 mL of nitric acid (14.2 mol/L) were placed in each digestion tube and capped with a cold finger. Groups of 30 tubes were heated to a temperature set at 150°C. Before placing the samples in the digestion block, pre-digestion was carried out for 30 minutes in the exhaust hood to avoid the formation of foam during the process. The aerial part was digested for around 2 hours, while the roots were digested for around 4 hours. After complete digestion, 2 mL of hydrogen peroxide was added to the material in two stages at 60-minute intervals. After 60 minutes of the 2^a addition of peroxide, the digester was switched off and the tubes were allowed to cool down before being filled into 15 mL falcon tubes, volumised to 10 mL and taken to the flame atomic absorption spectrometer. To read the samples, an analytical curve was determined with 5 standard points for Pb (1, 2, 3, 4 and 5 mg/L) and Cd (0.1; 0.2; 0.3; 0.4 and 0.5 mg/L) in a 0.5% nitric acid solution, according to Table 1.

Table 1 - Standard analytical curve used to determine Cd and Pb levels.

Standards mg/L				
Heavy metals	**P*1**	**P2**	**P3**	**P4**	**P5**
Cd	0.1 mg/L	0.2 mg/L	0.3 mg/L	0.4 mg/L	0.5 mg/L
Pb	1 mg/L	2 mg/L	3 mg/L	4 mg/L	5 mg/L
Standards pL				
Heavy metals	**P1**	**P2**	**P3**	**P4**	**P5**
Cd	5 μL	10 μL	15 μL	20 μL	50 μL
Pb	50 μL	100 μL	150 μL	200 μL	500 μL

*Standard.

Figure 1 - Equipment and samples used to analyse cadmium and lead in castor bean lines subjected to contaminated soil in Santo Amaro - Bahia. A: contrAA 700 flame atomic absorption spectrometer; B: Digester block and C: samples ground for digestion.

The data obtained was subjected to analysis of variance and the means were compared using the Tukey test at a 5% probability level, using the GENES statistical programme, version 2008.

RESULTS AND DISCUSSION

Bearing in mind that the complete castor bean cycle generally takes between 150 and 200 days (Silva et al. 2009), the 120-month phenological development of the strains submitted to the treatments in this study was generally satisfactory from the point of view of obtaining material for the analyses. With the exception of the strains/treatments (L1T1; L3T1; L4T1; L1T2; L1T3 and L1T4), all developed at the flowering stage. However, some plants showed chlorosis due to the presence of the trace element cadmium, slowing their growth compared to the others (Figure 2).

Costa et al. (2011) found that Cd caused severe phytotoxicity damage with visual symptoms throughout the plant when they carried out experiments that tested the phytoremediation capacity of the common castor bean to extract cadmium from soil treated with sewage sludge. The work carried out by Kupper et al. (2004) revealed that the chlorosis of castor bean leaves reflects the degree of cadmium toxicity present in the soil.

Figure 2 - Plants with chlorosis due to the presence of the trace element cadmium in soil from Santo Amaro - Bahia.

The analysis of variance showed that there was no significant difference in the elements cadmium and lead in the aerial part for the strains L1 (26); L2 (75); L3 (98) except for the strains L4 (175) and L5 (254) as shown in Table 2. This shows that the treatments (less contaminated soil, contaminated soil, contaminated soil + manure and contaminated soil + chelating agent) did not influence the stem and leaf development of the plants in 3 strains. On the other hand, there was a significant difference in the elements cadmium and lead in the roots, except in strains L1 (Cadmium in the root) and L4 (Lead in the root).

Table 2 - Summary of Analysis of Variance related to Cadmium in the aerial part (CdA), Cadmium in the root (CdR), Lead in the aerial part (PbA) and Lead in the root (PbR) in five castor bean lines in contaminated soil in Santo Amaro, Bahia.

Lineages	FV	GL	QM			
			CdA	CdR	PbA	PbR
L1	Treatment.	3	0.0041^{ns}	0.0111^{ns}	0.2869^{ns}	4,37*
	Res.	12	0,0035	0,0051	0,2777	1,134
	CV(%)	-	219,33	104,93	214,86	87,41
L2	Treatment.	3	0.0041^{ns}	0,76*	29.731^{ns}	18,034*
	Res.	12	0,0048	0,0134	20,34	2,68
	CV(%)	-	172,70	71,15	174,38	71,37
	Treatment.	3	0.0014^{ns}	0,284*	0.1573^{ns}	6,90*

L3	Res.	12	0,009	0,0039	0,1301	1,148
	CV(%)	-	161,43	61,13	223,85	61,92
L4	Treatment.	3	0,0003*	0,0235*	6.0773 ns	7.7020 ns
	Res.	12	0,00008	0,0067	6,8043	2,6502
	CV(%)	-	21,63	69,32	334,56	83,90
L5	Treatment.*	3	0,00012	0,0423*	16.393 ns	18,39*
	Res.	12	0,00003	0,006	17,181	1,10
	CV(%)	-	37,84	54,49	365,21	34,24

Captions: * (significant by F test at 5% probability level); [ns] (not significant); L1 (Lineage 26), L2 (Lineage 75), L3 (Lineage 98), L4 (Lineage 175) and L5 (Lineage 254).

According to the results, all the strains studied were successful in absorbing cadmium and lead in both the aerial parts and the roots, in all four treatments (T1, T2, T3 and T4). However, Table 3 shows that there was better absorption of lead in the aerial part for all treatments by all strains and that Lineage 175 (L4) was the best at absorbing both heavy metals, followed by Lineage 26 (L1).

The variation in the absorption of Cd in the roots compared to the aerial parts, compared to the absorption of Pb, may be related to the limitation of translocation, elucidating the complexation of these heavy metals with chelatins present in the root cells (HONG et al. 2008). Shortly after this process, these metals are sequestered by the central vacuoles and immobilised up to the threshold. Accioly et al. (2004) working with eucalyptus seedlings (Eucalyptus camaldulensis) found higher levels of Cd in the roots compared to the aerial part, indicating a limitation in the translocation of this metal. Chandra et al. (2010) investigating the distribution and bioaccumulation of Cd and Cr in pigeonpea (Vigna unguiculata), found a higher level of Cd deposition in the roots associated with its possible reaction with organic substances present in the secretions resulting from trauma; preventing its movement to the aerial part of the plant, probably the same occurred with the castor bean lines used in this study in comparison to the absorption of the heavy metal Pb.

This study found greater absorption of Pb in the aerial part when compared to the root. Discrepant data was found by Marques (2009) who worked with plants in the phytoextraction of lead, including castor beans, in contaminated areas around the automotive accumulator recycling plant in the municipality of Rio Tinto in Paraíba. He cultivated castor bean trees in clumps and collected samples 30 and 60 days after transplanting. By digesting and quantifying the metal content using atomic absorption spectrophotometry, he found that phytoaccumulation in the aerial part was not representative in relation to the roots. He points out that castor bean has a high capacity for absorbing and accumulating Pb in its root system, offering resistance to translocation to the aerial part, making the species suitable as a phytostabiliser. However, in this study, there was also good absorption of Pb by the roots, except when the strains were subjected to treatment T1.

Santos (2012), researching the accumulation of Pb in the plant tissue of papaya trees in contaminated soil to check their potential for phytoremediation, cultivated plants in pots, harvesting them 90 days after sowing and then quantifying the dry matter of the aerial part and the roots. He found that the plants accumulated more Pb in the roots than in the aerial part and had a consequent reduction in growth.

As well as affecting the osmotic balance of the plant cell, Cd can also influence the absorption, movement and assimilation of elements such as Mg, K, P, Ca and Fe by plant structures, according to: DAS et al. 1997; SANITÁ and GABRIELLI, 1999 ZEITOUNI et al. 2007). One of the positive aspects of cadmium, unlike other toxic elements, lies in the fact that it is easily absorbed by plant roots, especially in acidic soils as cited by Clemens, 1999, in agreement with the data obtained in this study.

Costa et al. (2011) assessed the tolerance of castor bean to the accumulation of Cd and Pb for phytoremediation purposes. The plants were grown in a nutrient solution with increasing doses of Cd and Pb, as well as the chelating agent EDTA, measuring the dry matter production of the aerial part and roots. They found that Cd caused severe phytotoxicity damage with visual symptoms throughout the plant, which did not occur with Pb, confirming the idea that this plant is suitable as a phytoremediator for contaminated soils. The chelating agent did not influence the displacement of metals for the variety studied. However, Khan et al. (2000) reported that due to their great ability to complex trace metals, chelating agents are used to promote the availability of these elements in contaminated soils. The effects of phytoextraction are greatly increased with the use of synthetic chelates such as ethylenediaminetriacetic acid (EDTA).

Table 3 - Cadmium and lead absorption in the aerial part (CdA and PbA) and in the root (CdR and PbR) of five castor bean lines as a function of 4 treatments in contaminated soil in Santo Amaro, Bahia.

VARIABLE	TREATS .	LINES				
		L1 (26)	L2 (75)	L3 (98)	L4 (175)	L5 (254)
CdA	T1	0,001 a	0,007 a	0,014 a	0,135 b	0,009 b
	T2	0,014 a	0,021 a	0,453 a	0,166 a	0.014 ab
	T3	0,073 a	0,078 a	0,148 a	0,171 a	0,228 a
	T4	0,019 a	0,054 a	0,128 a	0,184 a	0.015 ab
CdR	T1	0,002 a	0,006 b	0,019 b	0,003 a	0,007 b
	T2	0,099 a	0,299 a	0,148 a	0,163 a	0,251 a
	T3	0,050 a	0.229 ab	0,190 a	0,143 a	0.125 ab
	T4	0,120 a	0,065 b	0.066 ab	0,161 a	0,176 a
PbA	T1	0,002 a	0,143 a	0,001 a	0,005 a	0,005 a
	T2	0,028 a	6,277 a	0,451 a	2,621 a	0,038 a
	T3	0,439 a	1,179 a	0,106 a	0,263 a	0,334 a
	T4	0,512 a	2,876 a	0,087 a	0,230 a	4,164 a
PbR	T1	0,001 b	0,232 b	0,027 b	0,006 a	0,149 b
	T2	1,385 ab	4,129 a	2,837 a	2,801 a	5,093 a
	T3	0.962 ab	4,078 a	2,688 a	1,877 a	4,132 a
	T4	2,523 a	0.947 ab	1,369 ab	3,072 a	2,892 a

Captions: Treatments: T1 (less contaminated soil); T2 (contaminated soil); T3 (contaminated soil + organic matter) and T4 (contaminated soil + chelating agent). Strains: L1 (strain 26); L2 (strain 75); L3 (strain 98); L4 (strain 175) and L5 (strain 254). Averages followed by the same letter in the row and column do not differ statistically by the Tukey test at a 5% probability level.

CONCLUSION

All the strains tested were successful in absorbing cadmium and lead, with Strains 175 and 26 standing out, making them suitable for phytoextracting cadmium and lead from soils contaminated by these elements in Santo Amaro, Bahia.

BIBLIOGRAPHICAL REFERENCES

ACCIOLY, A. M. A.; SIQUEIRA, J. O; CURI, N.; MOREIRA, F. M. S. Limestone amelioration of zinc and cadmium toxicity for Eucalyptus camaldulensis seedlings grown in contaminated soil. Brazilian Journal of Soil Science. Viçosa, v. 28, n. 4, p. 775-783, jul./ago. 2004.

ANDRADE, J. C. M.; TAVARES, S. L. R.; MAHLER, C. F. Phytoremediation: the use of plants to improve environmental quality. São Paulo: Oficina de textos, 2007. 176 p.

ATSDR - AGENCY FOR TOXIC SUBSTANCES AND DISEASE REGISTRY: U.S. DEPARTMENT OF HEALTH AND HUMAN SERVICES. Draft toxicological profile for perfluoroalkyls. 2009. Available at: < http:// www.atsdr.cdc.gov/cercla/05list.html>. Accessed: 04 Apr. 2013.

CHANDRA, R. P.; ABDUSSALAM A. K.; SALIM, N. Distribution of bio-accumulated Cd and Cr in two Vigna species and the associated histological variations. Journal of Stress Physiology & Biochemistry. v. 6, n. 1, p. 4-12. 2010.

CHEN, S.; SUN, L.; SUN, T.; CHAO, L.; GUO, G. Interaction between cadmium, lead and potassium fertilizer (K_2SO_4) in a soil-plant system. Environmental Geochemistry and Health, Amsterdam, v. 29, n. 5, p. 435-446, Apr. 2007.

CLEMENS, S. Molecular mechanisms of plant metal tolerance and homeostasis. Planta, v. 212, p. 475-486, sep. 2000.

COSTA, E. T. S.; GUILHERME, L. R. G.; MELO, E. E. C.; RIBEIRO, B. T.; INÁCIO, E. S. B.; SEVERIANO, E. C.; FAQUIN, V.; HALE, B. A. Assessing the Tolerance of Castor Bean to Cd and Pb for Phytoremediation Purposes. Springer Science + Business Media. p. 1-3, Jul. 2011.

DAS, P.; SAMANTARAY, S.; ROUT, G. R. Studies on cadmium toxicity in plants: a review. Environmental Pollution, v. 98, n. 1, p. 29-36. 1997.

FREITAS, E. V. S.; NASCIMENTO, C. W. A.; SILVA, A. J.; DUDA, G. P. Induction of lead phytoextraction by citric acid in soil contaminated by automotive batteries. Revista Brasileira de Ciência do Solo, Viçosa, v. 33, n. 2, p. 467-473, mar/abr. 2009.

GOPAL, R.; RIZVI, A. H. Excess lead alters growth, metabolism and translocation of certain nutrients in radish. Chemosphore, v. 70, n. 9, p. 15391544, feb. 2008.

GUIMARÃES, M. A.; SANTANA, T. A.; SILVA, E. V.; ZENZEN, I. L. LOUREIRO, M. E. Cadmium toxicity and tolerance in plants. Revista Tropica - Ciências Agrárias e Biológicas, v. 1, n. 3, p. 58, 2008.

HONG, C. L.; JIA, Y. B.; YANG , X. E.; HE, Z. L.; STOFFELLA, P. J. Assessing lead thresholds for phytotoxicity and potential dietary toxicity in selected vegetable crops. Bull. Environ. Contam. Toxicol, v. 80, p. 356-361, 3 Mar. 2008.

ISHIKAWA, D. N.; NOALE, L. Z.; OHE, T. H. K.; SOUZA, E. B. R.; SCARMINIO, I. S.; BARRETO, W. J.; BARRETO, S. R. G. Evaluation of the environmental risk in sediments of the Cambé creek lakes, in Londrina, by the distribution of metals. Química Nova, v.32, n. 7, p. 1744-1749, 2009.

KABATA-PENDIAS, A.; PENDIAS, H. Trace elements in soils and plants. 3rd ed. Boca Raton, USA: CRC Press. 2000, 413 p.

KHAN, A. G.; KUEK, C.; CHAUDHRY, T.M.; KHOO, C. S.; HAYES, W. J. Role of plants, mycorrhizae and phytochelators in heavy metal contaminated land remediation. Chemosphere, v. 41, n. 1, p. 197-207, 2000.

KUPPER, H.; MIJOVILOVICH, A.; MEYERKLAUCKE, W.; KRONECK, M. H. Tissueand age-dependent differences in the complexation of cadmium and zinc in the cadmium/zinc hyperaccumulator Thlaspi caerulescens (Ganges Ecotype) revealed by Xray absorption spectroscopy. Plant Physiology. v. 134, n. 2, p. 748-757, feb. 2004.

LIMA, A. M. Evaluation of the Phytoremediation Potential of Castor Bean (Ricinus communis L.) and Sunflower (Helianthus annus L.) for the Removal of Lead and Toluene from Synthetic Effluents. 2010. 110 f. Thesis (Doctorate in Chemical Engineering) - Federal University of Rio do Norte. Natal, 2010.

LIMA, F. S. Bioconcentration of lead and zinc in edible parts of vegetables grown in contaminated soils. 2010. 89 f. Thesis (Doctorate in Soil Science) - Federal Rural University of Pernambuco, Recife, 2010.

LINHARES, L. A.; EGREJA FILHO, F. B.; OLIVEIRA, C. V.; BELLIS, V. M. Adsorption of cadmium and lead in highly weathered tropical soils. Pesquisa Agropecuária Brasileira, Brasília, DF, v. 44, n. 3, p. 291-299, mar. 2009.

LU, A.; ZHANG, S. and SHAN, X-Q. Time effect on the fractionation of heavy metals in soils. Geoderma, v. 125, n. 3, p. 225-234, Apr. 2005.

MANECKI, M; BOGUCKA, A; BADJA, T; BORKIEWICZ, O. Decrease of Pb bioavailability in soils by addition of phosphate ions. Environmental Chemistry Letters, v. 3, n. 4, p. 178-181, jan. 2006.

MARQUS, L. F. Lead phytoextraction by sunflower, vetiver, buckwheat, jureminha and castor bean in contaminated areas. 2009. 50 f. Dissertation (Master's Degree in Agronomy) - Centre for Agrarian Sciences, Federal University of Paraíba, Areia, 2009.

MELO, E. E. C.; NASCIMENTO, C. W. A.; SANTOS, A. C. Q.; SILVA, A. S. Availability and fractionation of Cd, Pb, Cu and Zn as a function of pH and soil incubation time. Ciência Agrotécnica, Lavras, v. 32, n. 3, p. 776-784, mai/jun. 2008.

PIECHALAK, A.; TOMASZEWSKA, B.; BARALKIEWICZ, D.; MALECKA, A. Accumulation and detoxification of lead ions in legumes. Phytochemistry, v. 60, n. 2, p. 153-162, 2002.

SANITÁ DI TOPPI, L.; GABBRIELLI, R. Response to cadmium in higher plants. Environmental and Experimental Botany, v. 41, n. 2, p.105-130, abr. 1999.

SANTOS, C. H. S. Potential for Pb phytoextraction by papaya trees in contaminated soil. Semina, Londrina, v. 33, n. 4, p. 1427-1434, jul./ago. 2012.

SILVA, J. J. N.; MONTENEGRO, A. A. A; BELTRÃO, N. E. M.; CARTAXO, W. V.; OLIVEIRA, M. X. Growth and productivity of castor bean (Ricinus *comunnis* L.) in family farming in the semi-arid region of Pernambuco. In: 7° SIMPÓSIO BRASILEIRO DE CAPTAÇÃO E MANEJO DE ÁGUA DA CHUVA. 2009. Caruaru, ANAIS...

TORRI, S.; LAVADO, R. Plant absorption of trace elements in sludge amended soils and correlation with soil chemical speciation. Journal of Hazardous Materials, Amsterdam, v. 166, p. 1459-1465, Jul. 2009.

ZEITOUNI, C. F.; BERTON, R. S.; ABREU, C. A. Phytoextraction of cadmium and zinc from a red-yellow latosol contaminated with heavy metals. Bragantia, Campinas, v. 66, n. 4, p. 649-657, 2007.

FINAL CONSIDERATIONS

The castor bean plant has topped the list of plant species that can satisfactorily absorb considerable quantities of toxic substances without interrupting their vital activities. Some castor bean strains manage to extract different quantities when chelating agents such as EDTA are added. The contributions of this work to the

Santamarense community will be of great relevance, since that population suffers the consequences of environmental mismanagement which has caused such damage over more than 20 years. Information on the phytoextraction potential of the strains developed by the NBIO/UFRB could minimise the pollution resulting from the activities of the former Companhia Brasileira de Chumbo in the medium or long term, by planting them in the most critical areas related to contamination. With the possibility of extracting cadmium and lead residues, the seeds could also be used to produce biofuel, since there is no record of these metals accumulating in the fruit.

I want morebooks!

Buy your books fast and straightforward online - at one of world's fastest growing online book stores! Environmentally sound due to Print-on-Demand technologies.

Buy your books online at
www.morebooks.shop

Kaufen Sie Ihre Bücher schnell und unkompliziert online – auf einer der am schnellsten wachsenden Buchhandelsplattformen weltweit! Dank Print-On-Demand umwelt- und ressourcenschonend produziert.

Bücher schneller online kaufen
www.morebooks.shop

info@omniscriptum.com
www.omniscriptum.com